LONDON MIDLAND STEAM

ABOUT THE PHOTOGRAPHER

R.J. (Ron) Buckley was born in 1917, and after the family moved to a house overlooking Spring Road station, near Tyseley, in 1926 his interest in railways grew. Joining the Birmingham Locomotive Club in 1932, he made frequent trips with them throughout the country, also accompanying W.A. Camwell on his many branch-line tours. From 1936 until 1939 these club tours included visits to Manchester, Lancashire, Cumberland and Westmorland where he photographed many examples of the pre-grouping locomotive classes still working with the London, Midland and Scottish Railway (LMS).

In 1934 he joined the LMS as a wages clerk at Lawley Street Goods station, Birmingham and after the declaration of war in September 1939 was called up and joined the Royal Engineers, being posted to the No. 4 Dock Operating Unit. Serving briefly in Norway during 1940, he was by May of that year with a special party supplying stores for the returning troops at Dunkirk. Ron was evacuated from Dunkirk on the *Maid of Orleans,* an ex-Southern Railway cross-Channel ferry,

and by 1941 he was in Egypt with his unit supporting the 8th Army in its advance from Alamein, finally reaching Tripoli. The year 1944 saw his unit in Alexandria before returning to Britain during 1945 and demobilisation in May 1946.

His Employment continued with the LMS being based at Kings Heath in Birmingham and at Derby in the British Railways Divisional Manager's Office from 1948. He was a spectator of the continual stream of locomotives passing through the works and photographed many of the new British Railways Standard locomotives constructed there. Other organised visits included both Crewe and Horwich Works and throughout the 1950s and early 1960s he witnessed the many changes that were taking place in locomotive power throughout the old LMS territory.

Married in 1948 to Joyce, the daughter of an LNER locomotive driver, Ron retired in 1977 after over forty-two years' service with the railways. He and his wife currently live in Staffordshire.

PHOTOGRAPHS

The photographs reproduced in the six published volumes of Ron Buckley's books are only a small selection of the more than 4,000 negatives in his catalogue of work. This complete catalogue can be obtained by writing to Mr Colin Stacey at Initial Photographics, 25 The Limes, Stony Stratford, MK11 1ET.

LONDON MIDLAND STEAM

THE RAILWAY PHOTOGRAPHS OF R.J. (RON) BUCKLEY

COMPILED BY BRIAN J. DICKSON

The
History
Press

Other books in this series:

Steam in Scotland: *The Railway Photographs of R.J. (Ron) Buckley*

Southern Steam: *The Railway Photographs of R.J. (Ron) Buckley*

Steam in the North East: *The Railway Photographs of R.J. (Ron) Buckley*

Great Western Steam: *The Railway Photographs of R.J. (Ron) Buckley*

Steam in the East Midlands and East Anglia: *The Railway Photographs of R.J. (Ron) Buckley*

First published 2018

The History Press
The Mill, Brimscombe Port
Stroud, Gloucestershire, GL5 2QG
www.thehistorypress.co.uk

© Brian J. Dickson, 2018

The right of Brian J. Dickson to be identified as the Author
of this work has been asserted in accordance with the
Copyright, Designs and Patents Act 1988.

British Library Cataloguing in Publication Data.
A catalogue record for this book is available from the British Library.

ISBN 978 0 7509 8796 7

Typesetting and origination by The History Press
Printed in Turkey by Imak

Front cover: Saturday 28 July 1951. Ex-LMS Class 5 2-6-0 No 42818 is awaiting her next duty at Birmingham New Street station.

Back cover: Monday 26 April 1954. The driver of ex-LMS Class 5XP 4-6-0 'Jubilee' No 45610 *Gold Coast* has applied the steam sanders to aid his departure from Derby Midland station at the head of the 12.40 p.m. Newcastle to Bristol express. Constructed at Crewe Works during 1934, she would be withdrawn in 1964. During 1958 she would be renamed *Ghana* after the Gold Coast had gained independence from the United Kingdom.

Half title: Thursday, 1 June 1939. With a crew member overseeing the filling of the tender with water, ex-L&YR Class 27 (LMS Class 3F) 0-6-0 No. 12156 is ready for duty at Bacup shed. Constructed at Horwich Works during 1892, she would be numbered 52156 by British Railways before being withdrawn in 1953.

Title page: Thursday, 3 March 1938. At New Street station in Birmingham, LMS Class 5XP 4-6-0 'Jubilee' No. 5606 *Falkland Islands* is looking splendid in her crimson lake livery. The diamond-shaped works plate on the front frame shows that she was constructed by the North British Locomotive Works in Glasgow. Entering service during 1935, she would be withdrawn in 1964.

INTRODUCTION

This volume of Ron Buckley's photographs focuses on the London Midland and Scottish Railway (LMS) and its pre-grouping constituent company locomotives working in England and Wales.

With the grouping of the railways in January 1923, the two largest constituent companies forming the LMS were the London North Western Railway (LNWR) and the Midland Railway (MR), both of which operated main lines from London that connected at Carlisle with their Scottish partners. The LNWR had, for many years, relied on a series of 4-4-0 locomotives for their express passenger services, but with escalating loads the 4-6-0 wheel arrangement became paramount in their ability to handle the increasing demands in traffic. From 1905 a series of impressive two-cylinder 4-6-0s appeared, culminating in the four-cylinder superheated 'Claughtons', which were introduced during 1913 with the last entering service in 1921. With a total of over 600 of these 4-6-0s in the LNWR locomotive fleet, many continued in traffic with the LMS until the late 1930s and early 1940s,with the last being withdrawn in February 1950.

The Midland Railway also relied on a series of very able 4-4-0s for express passenger traffic, utilising both the two-cylinder simple and three-cylinder compound versions. So competent were these locomotives that the vast majority led extended lives in traffic after being fitted with superheating boilers. Several hundred examples passed into the hands of the LMS at the grouping and, with the strong Derby 'small engine' influence within the LMS immediately after the grouping, a further 195 Class 4P three-cylinder compounds were constructed between 1924 and 1932. In addition, another 138 examples of the Class 2P two-cylindered 4-4-0s entered service from 1928, with the last appearing in 1932. Being allocated as widely as Aberdeen and Ayr in Scotland to Bath and Templecombe on the Somerset and Dorset Joint Railway (S&DJR), the bulk of the

4P compounds survived until the late 1950s, with many of the 2Ps going for scrap in the early 1960s.

As a smaller constituent of the LMS, the Lancashire and Yorkshire Railway (L&YR) had actually amalgamated with the LNWR during 1922, being primarily a goods and mineral carrying company. The passenger services between the major cities and towns within its sphere of influence were again handled by a series of two-cylinder 4-4-0s constructed between 1888 and 1915 and numbered about seventy examples. In addition, a single class of 'Atlantics', with enormous 7ft 3in driving wheels, was produced between 1899 and 1902 and totalled forty examples. It was not until 1908 that the L&YR saw its only class of locomotives with a 4-6-0 wheel arrangement enter service, with the four-cylinder, un-superheated-boiler Class 8 extending to twenty examples. These proved to be unreliable in service and all were rebuilt during 1920 and 1921 with new cylinders, Walschaerts valve gear and superheated boilers, which improved their performance markedly. A further fifty-five examples were constructed between 1921 and 1925 with the bulk being withdrawn during the 1930s, a few surviving into the late 1940s and the last example being withdrawn in 1951.

The early years of the LMS saw the ex-L&YR Chief Mechanical Engineer (CME) George Hughes installed as its first CME, but after his early retirement the replacement was the former Midland Railway CME Henry Fowler, later Sir Henry, as he was knighted in 1920 for his services to the war effort. By the mid-1920s there was an urgent need for new express passenger locomotives and in 1927 an order was placed with the North British Locomotive Co. (NBL) in Glasgow for a three-cylinder 4-6-0, with both Derby and the NBL drawing offices working together on the final design. With its huge manufacturing capability, the NBL delivered fifty class members between September and December of that same year. Thus the highly successful 'Royal Scot' class was born, with a further twenty examples being constructed at Derby Works during 1930. What ensued was a plan by Henry Fowler to rebuild the ageing 'Claughtons' into a three-cylinder version with similarities to the 'Royal Scots', but in reality these were new builds only using some parts from the 'Claughtons'. A total of fifty-two locomotives emerged between 1930 and 1934 in what became known as the 'Patriot' class.

After the retirement of Sir Henry Fowler in 1931, it was a year later before William Stanier stepped into the office of Chief Mechanical Engineer and so began a period of exceptional locomotive design that brought forth within five years two classes of four-cylinder 4-6-2 express passenger locomotives, one class of three-cylinder 4-6-0 express passenger locomotives, one class each of a two-cylinder 2-6-0 and 4-6-0 mixed traffic locomotives, three tank locomotive classes and a 2-8-0 two-cylinder goods locomotive. The first class of 'Pacifics' entered service in July 1933 with the 'Princess Royal' series of twelve locomotives coming out of Crewe Works, the final example entering service in 1935. Concurrent with this event, the two-cylinder Class 5 2-6-0 mixed-traffic locomotives were also appearing in traffic from Crewe Works. Commencing in 1934, examples of the three-cylinder 'Jubilee' Class 5XP 4-6-0s were coming out of Crewe Works and the NBL with large numbers of the two-cylinder Class 5, or the 'Black 5s', being delivered from outside suppliers. During the same year Derby Works was busy delivering the new Stanier design of 2-6-4 tanks. The year 1935 saw further deliveries of both 2-6-2 and 2-6-4 tanks from Derby Works and the commencement from Crewe Works of the powerful two-cylinder Class 7F, later 8F 2-8-0 goods locomotives. It was 1937 before the first examples of what is generally described as Stanier's masterpiece design were delivered from Crewe Works in the form of his 'Princess Coronation'

four-cylinder 4-6-2s. The streamlined form of No. 6220 *Coronation* entered service in June of that year to be followed by a further thirty-seven examples over a period of eleven years, the last appearing in 1948. So successful were some of these designs that construction of the 'Black 5s' continued from 1934 until 1951 when a total of 842 examples had entered service from the three main LMS Works and two independent manufacturers. Similarly, the 8F was produced in prolific numbers with a total of 852 examples being constructed by three independent manufacturers and eight pre-nationalisation railway works – Ashford, Brighton, Crewe, Darlington, Doncaster, Eastleigh, Horwich and Swindon. With many being constructed for the War Department during the Second World War they found their way to such far-flung parts of the globe as Egypt and Turkey, with China being the furthest reached. William Stanier was knighted in 1943 for his services to engineering.

At grouping, the LMS found itself the owners of about 2,000 six-wheeled goods locomotives inherited from its English constituent companies, the bulk coming from the LNWR, L&YR and the MR. These, together with approximately 800 examples of the former LNWR and L&YR 0-8-0s, were the stalwarts in the movement of goods and mineral traffic over much of the LMS. In addition to these pre-grouping locomotives, the LMS constructed over 500 examples of the Henry Fowler design of 4F 0-6-0 throughout the 1920s. Ron Buckley's photographs show the changes in the locomotive scene that took place throughout the LMS territory, illustrating from the later 1930s those pre-grouping classes still working and the newer classes being introduced by them. Ron's later photographs, from 1946 onwards, show more of these remaining working pre-grouping locomotives but also portray the newer designs of Henry Fowler and William Stanier.

Tuesday, 19 May 1936. At Newton Heath shed in Manchester, two examples of the ex-L&YR Class 5 (LMS Class 2P) 2-4-2 tanks were in residence. *Right:* No. 10766 was constructed at Horwich Works during 1897 and numbered 1333 by the L&YR. She would become No. 50766 with British Railways and be withdrawn in 1951.

Tuesday, 19 May 1936. No. 10652 was an earlier 1892 example of the class that was numbered 1161 by the L&YR. She would receive a Belpaire firebox during 1913 and be numbered 50652 before being withdrawn in 1956. This numerous class, with 330 examples all originating from Horwich Works, were from a design by John Aspinall that was continued by Henry Hoy and George Hughes over a period of twenty-two years from 1889 until 1911.

Opposite top: **Tuesday, 19 May 1936.** Also seen in the yard at Newton Heath shed is ex-L&YR Class 30 (LMS Class 6F) 0-8-0 No. 12726 bearing a 25A Wakefield shed code. Designed by John Aspinall and introduced in 1900 after he had become General Manager at that company, this example was a 1902 product of Horwich Works that would be withdrawn in December 1936, seven months after this photograph. *Opposite bottom:* Also at Newton Heath shed and bearing the correct 26A shed code, is the impressive bulk of LMS Class 7F 0-8-0 No. 9654. She was designed by Henry Fowler as an updated version of the earlier LNWR 0-8-0s and was introduced in 1929, with a total of 175 examples being constructed at Crewe Works. Built in 1932, she would give only seventeen years of service before being withdrawn during 1949.

Above: Wednesday, 20 May 1936. Seen at Colne shed is ex-L&YR Class 27 (LMS Class 3F) 0-6-0 No. 12181. Designed by John Aspinall and introduced during 1889, this example was constructed in 1893 at Horwich Works and was numbered 1195 by them. Bearing a 24C Lostock Hall shed code, she would be withdrawn during 1949. The shed at Colne, opened by the L&YR during 1900, would be closed four months after this photograph was taken.

Sunday, 31 May 1936. Seen here in the yard at Bolton shed is an example of the George Hughes-designed 4-6-4 'Baltic Tank' for the L&YR. With all ten examples being constructed at Horwich Works and entering service during 1924, they were all numbered directly into LMS service. Classified 5P by that company, these powerful four-cylinder locomotives handled heavy suburban traffic around Manchester for most of their working lives, with No. 11117 being withdrawn during 1941.

Sunday, 31 May 1936. Also seen at Bolton shed is ex-L&YR Class 1 'Railmotor' No. 10600. Based on a design by George Hughes, she was constructed during 1906 and numbered 3 by the L&YR, one of eighteen examples that were constructed between 1906 and 1911 to handle light branch traffic. She would be withdrawn in 1947.

Sunday, 8 November 1936. In Derby Works yard, LMS Class 4P 2-6-4 tank No. 2467 has been wheeled out of the works after completion and is yet to enter service. Designed by William Stanier, the class saw 206 examples constructed at both Derby Works and the NBL in Glasgow. No. 2467 would become No. 42467 with British Railways and be withdrawn during 1961.

Sunday, 14 February 1937. Parked in Crewe South shed yard is LMS Class 3F 0-6-0 tank No. 7475. Constructed at the Vulcan Foundry during 1927, she would be numbered 47475 by British Railways and withdrawn in 1962.

Saturday, 10 April 1937. Seen here in Derby shed yard is another of William Stanier's 4P 2-6-4 tanks. No. 2489 is about to enter service, having exited from the works earlier in the same week. She would become No. 42489 with British Railways and be withdrawn during 1964. Note the differing style of lettering used compared to that of classmate No. 2467 on page 11.

Friday, 7 May 1937. At Workington shed, ex-Furness Railway (FR) Class 98 (LMS Class 2P) 0-6-2 tank No. 11628 is having its fire cleaned. A survivor of a class of ten locomotives designed by William Pettigrew and constructed for that railway during 1904, five examples from Nasmyth Wilson & Co. and five from the NBL in Glasgow, she was numbered 101 by the FR. No. 11628 was a Nasmyth Wilson product that would be withdrawn during 1946.

Friday, 7 May 1937. At Workington shed again is LMS Class 2P 4-4-0 No. 659. Constructed at Crewe Works during 1931, she would be numbered 40659, before being withdrawn thirty years later in 1961.

Friday, 7 May 1937. At this time Moor Row shed 12E near Whitehaven had an allocation of about a dozen ex-FR, L&YR and LNWR 0-6-0s that were utilised on the traffic from the quarries in the Cumberland hills. *Left:* Ex-FR Class 1 (LMS Class 3F) 0-6-0 No. 12499 entered service from the NBL in Glasgow during 1914 and be numbered 28 by that railway. She would give forty-three years of service, be renumbered 52499, and be ultimately withdrawn in 1957.

Friday, 7 May 1937. Classmate No. 12494 was also from the NBL in Glasgow but a year earlier, 1913, and numbered 1 by the Furness Railway. She would be withdrawn during 1956, after being renumbered 52494 with British Railways. Designed by William Pettigrew, a total of nineteen examples of the class were constructed between 1913 and 1920.

Friday, 7 May 1937. Also at Moor Row shed is ex-LNWR 'Coal Engine'
(LMS Class 2F) 0-6-0 No. 8303. Constructed at Crewe Works during 1892,
she would give fifty-four years of service, only to be withdrawn in 1946.

Sunday, 6 June 1937. Introduced during 1912, the superheated boiler version of the Charles Bowen-Cooke-designed Class G 0-8-0 became Class G1, with construction continuing until 1918. Many of the earlier Class G locomotives were rebuilt with superheating boilers over a period of years. Seen here at Royston shed, near Barnsley, is ex-LNWR Class G1 (LMS Class 7F) No. 9297, constructed at Crewe Works in 1918 and withdrawn during 1949.

Sunday, 25 July 1937. Designed by John Aspinall and introduced in 1889, the Class 5 (LMS Class 2P) 2-4-2 tanks for the L&YR continued construction until 1911, by which time a total of 330 examples had entered service. No. 10807 is seen here at Bolton shed: she was constructed at Horwich Works in 1898 and would become No. 50807 with British Railways before being withdrawn during 1955.

Sunday, 25 July 1937. Also seen at Bolton shed is ex-L&YR Class 23
(LMS Class 2F) 0-6-0 saddle tank No. 11439. Originally constructed by the
Vulcan Foundry during 1882 as an 0-6-0 Class 528 tender locomotive to a
design by William Barton Wright, she was rebuilt in 1896 by John Aspinall as
a saddle tank. Numbered 51439 by British Railways, she would be withdrawn
in 1956, having given a total of seventy-four years of service.

Sunday, 10 October 1937. Crewe South shed is host to ex-L&YR Class 27 (LMS Class 3F) 0-6-0 No. 12143, bearing a 4D Abergavenny shed code. Constructed at Horwich Works during 1891 to a design by John Aspinall, she would be numbered 52143 and withdrawn in 1957.

Sunday, 10 October 1937. Also seen in Crewe South shed yard is ex-LNWR 0-6-0 saddle tank No. 27484. Originally constructed as an 0-6-0 'Coal Engine' locomotive to a design by Francis Webb during 1891, she would be converted to this unusual style of saddle tank in 1906, be classified 2F by the LMS and be withdrawn during 1946.

Sunday, 13 March 1938. Seen standing on one of the turntable roads at Derby Shed is ex-MR Class 2 (LMS Class 2P) 4-4-0 No. 418. Constructed by Sharp Stewart & Co. during 1892, she would serve for sixty-five years before being withdrawn in 1957.

Sunday, 13 March 1938. Designed by Henry Fowler for the Midland Railway specifically to handle goods and mineral traffic on the Somerset and Dorset Joint Railway, (S&DJR) the Class 7F 2-8-0s were highly successful in that task. Seen here at Derby Works is No. 13804, constructed at Derby during 1914 and numbered 84 by the S&DJR. She would become No. 9674 and later 13804 with the LMS and finally 53804 with British Railways. She would be withdrawn in 1962.

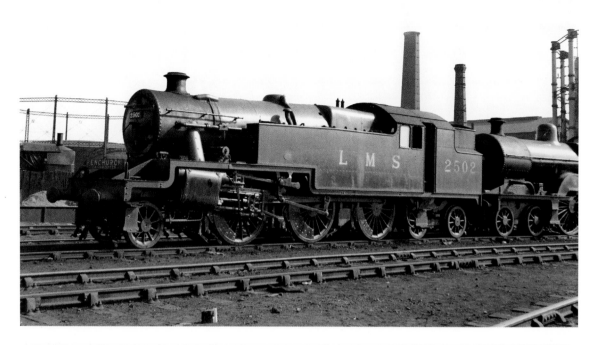

Sunday, 13 March 1938. Designed by William Stanier and introduced during 1934 specifically to work the tightly timed suburban traffic on the London, Tilbury and Southend route, his three-cylinder Class 4P 2-6-4 tanks, were all constructed at Derby Works in 1934. Seen here at Derby is No. 2502, which would become No. 42502 and be withdrawn during 1962.

Monday, 14 March 1938. Designed by Samuel Johnson for the Midland Railway (MR) and introduced during 1882 with construction continuing until 1901, the Class 2 4-4-0 express passenger locomotives were constructed at Derby Works but also by five outside contractors. No. 526, seen here at Derby shed, was a Derby Works product from 1898 that would be withdrawn during 1956.

Sunday, 10 April 1938. Standing in the yard at Mold Junction shed is ex-LNWR 0-6-2 'Coal Tank' (LMS Class 2F) No. 7757. Constructed at Crewe Works during 1886, she would be fitted with the LMS vacuum control system for motor train working in 1946 and be withdrawn from service during 1954, numbered 58915.

Sunday, 10 April 1938. Ex-LNWR Class G1 (LMS Class 7F) 0-8-0 No. 9258 is seen at Mold Junction shed. Constructed during 1916 at Crewe Works, she would undergo rebuilding in 1941 to a Class G2A, incorporating a higher working boiler pressure of 175psi, and would be withdrawn during 1951.

Tuesday, 21 June 1938. Ex-L&YR Class 5 (LMS Class 2P) 2-4-2 tank No. 10833 was a long-time resident of Blackpool Central shed, where she is seen here. Constructed at Horwich Works in 1898, she would be withdrawn prior to the nationalisation of the railways.

Tuesday, 21 June 1938. Also seen at Blackpool Central shed is LMS Class 5XP 4-6-0 'Jubilee' No. 5571 *South Africa*. An example of the fifty locomotives of the class constructed by the NBL in Glasgow during 1934, she would be withdrawn thirty years later in 1964.

Tuesday, 21 June 1938. By this date there were only ten surviving examples of George Hughes' four-cylinder Class 8 4-6-0s designed for express passenger traffic with the L&YR, most allocated to Blackpool Central shed. A total of fifty-five were constructed at Horwich Works from 1908, with the last appearing during 1924. Reputedly heavy coal users, the earlier locomotives were rebuilt with new cylinders and Walschaerts valve gear which resulted in an improved performance. *Opposite top:* Seen at Blackpool Central shed is No. 10429. Constructed in 1922, she would be withdrawn during 1948. *Opposite bottom:* Seen here on the turntable at Blackpool Central shed is No. 10448. Constructed in 1923 she would be withdrawn during 1949.

Above: Ex-L&YR Class 27 (LMS Class 3F) 0-6-0 No. 12415 is seen parked in the yard at Blackpool South shed. Constructed during 1900 at Horwich Works, she would be numbered 432 by the L&YR and become No. 52415 with British Railways before being withdrawn in 1961.

A comparison of two examples of the Samuel Johnson design of Class 1121 (LMS 1F) 0-6-0 tank for the Midland Railway.

Sunday, 17 July 1938. No. 1857 is seen at Sheffield in its original form. Constructed by Sharp Stewart & Co. during 1895, she would receive a Belpaire firebox during 1955 and be withdrawn in 1959, numbered 41857.

Sunday, 5 February 1939. A rebuilt No. 1878 is seen here at Gloucester shed. Constructed by Robert Stephenson & Hawthorn in 1899, she received the Belpaire firebox during 1929. Numbered 41878 by British Railways, she would be withdrawn in 1959.

Sunday, 5 February 1939. Designed by Henry Fowler for the LMS as a multipurpose locomotive, the Class 3F 0-6-0 tanks were introduced during 1924. Construction continued until 1931 with 422 examples entering service. Seen here at Gloucester shed is No. 7275, a product of the Vulcan Foundry in 1924 that would become No. 47275 and be withdrawn during 1962.

Sunday, 14 May 1939. At Lostock Hall shed, the driver of ex-L&YR Class 5 (LMS Class 2P) 2-4-2 tank No. 10823 is preparing the locomotive for duty. Constructed at Horwich Works during 1898, she was rebuilt with a Belpaire firebox, as seen here, during 1912 and would be withdrawn in 1948.

Sunday, 14 May 1939. Also at Lostock Hall is ex-L&YR Class 28 (LMS Class 3F) 0-6-0 No. 12606. Originally constructed at Horwich Works in 1889 with a saturated boiler, she was rebuilt during 1922 with a Belpaire firebox and superheating boiler. Destined to be requisitioned by the Railway Operating Division (ROD) in 1917, she would be returned to the L&YR during 1919 and withdrawn from service in 1946.

Thursday, 1 June 1939. Seen here at Bury shed are sister 0-6-0 locomotives from ex-L&YR Class 28 (LMS Class 3F). *Left:* No. 12580, constructed at Horwich Works in 1889 and subsequently rebuilt with a superheating boiler and Belpaire firebox during 1919, would become No. 52580 with British Railways and be withdrawn in 1954.

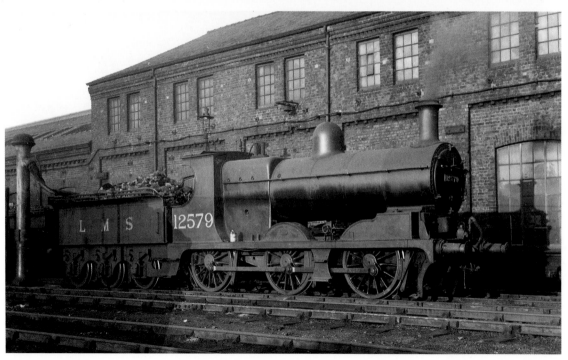

Thursday, 1 June 1939. No. 12579 was also a product of Horwich Works, but from 1896. She would be rebuilt during 1916 with a Belpaire firebox and superheating boiler, before becoming No. 52579 and being withdrawn in 1952.

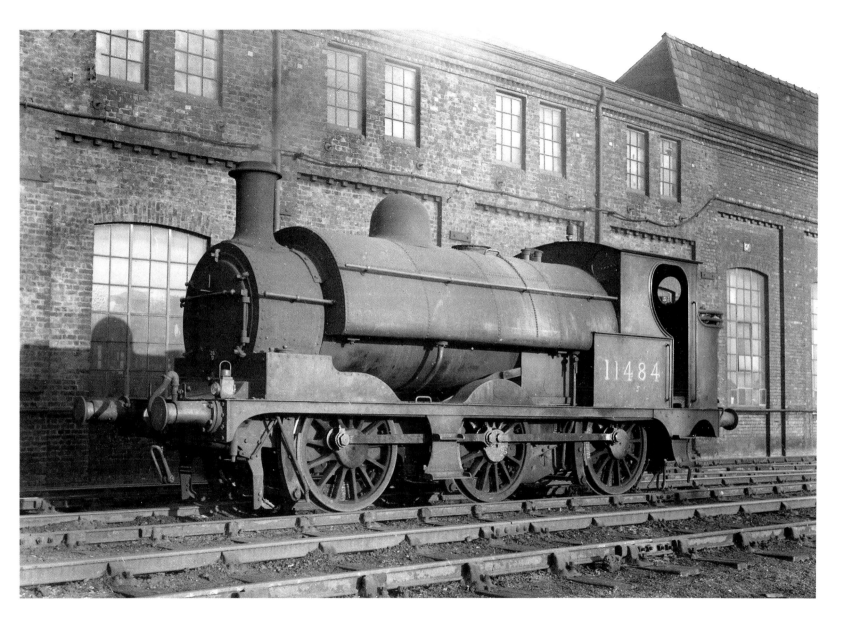

Thursday, 1 June 1939. Ex-L&YR Class 23 (LMS Class 2F) 0-6-0 saddle tank
No. 11484 is also seen at Bury shed. Originally designed as an 0-6-0 tender
locomotive and constructed during 1882 by Beyer Peacock & Co., she would
be rebuilt as a saddle tank during John Aspinall's period as Locomotive
Superintendent in 1898. Becoming No. 51484 with British Railways, she
would be withdrawn in 1959.

Sunday, 13 August 1939. This spread shows a trio of ex-LNWR 'Cauliflower' Class (LMS Class 2F) 0-6-0s all photographed on the same day. *Above:* Seen at Workington shed is No. 8515. Constructed at Crewe Works during 1900 she would become No. 58398 with British Railways and be withdrawn in 1952.

Also at Workington shed is No. 8415, another Crewe Works product but of earlier construction, built during 1896. Not reaching British Railways stock, she was withdrawn in 1948.

Seen at Upperby shed in Carlisle is No. 8369. An 1895 product of Crewe Works, she would be withdrawn from service during 1947.

Sunday, 13 August 1939. The 'Patriot' 5XP class of three-cylinder 4-6-0s were originally conceived by Henry Fowler as rebuilds of the ex-LNWR 'Claughton' class locomotives with a total of fifty-two examples being constructed. No. 5537 *Private E. Sykes V.C.*, seen here at Barrow-in-Furness shed, was constructed at Crewe Works during 1933. Several examples of the class were rebuilt with Stanier-designed taper boilers, but No. 5537 was not chosen for rebuilding and was withdrawn from service in the form seen here in 1962.

Monday, 14 August 1939. Ex-LNWR Class G2A (LMS Class 7F) 0-8-0 No. 9238 is seen at Oxenhope shed. Constructed at Crewe Works during 1914 as a Class G1 locomotive, she had been rebuilt as a Class G2A earlier in 1939 and would be withdrawn during 1949.

Wednesday, 28 May 1947. At Coaley Junction ex-MR Class 1377 (LMS Class 1F) 0-6-0 tank No. 1720 is seen waiting to depart with the 8.00 p.m. working to Dursley. Constructed at Derby Works during 1883, she would be rebuilt in 1926 with a Belpaire firebox and withdrawn during 1956, numbered 41720. Coaley Junction Station opened in September 1856, and survived until complete closure in September 1962.

Saturday, 27 September 1947. The simple one-road shed at Knighton in rural Shropshire was a sub-shed of Shrewsbury, shed code 4A, and here it is paying host to ex-MR Class 2 (LMS Class 3F) 0-6-0 No. 3679. Constructed by Kitson & Co. during 1901, she would be withdrawn in 1962, after being numbered 43679 by British Railways.

Sunday, 7 March 1948. One of the Crewe Works shunters was ex-LNWR John Ramsbottom-designed 0-6-0 'Special tank' (LMS Class 2F) No. 3323, which had been constructed at the same works in 1878. She would continue in that role until she was withdrawn during 1954, having given seventy-six years of service.

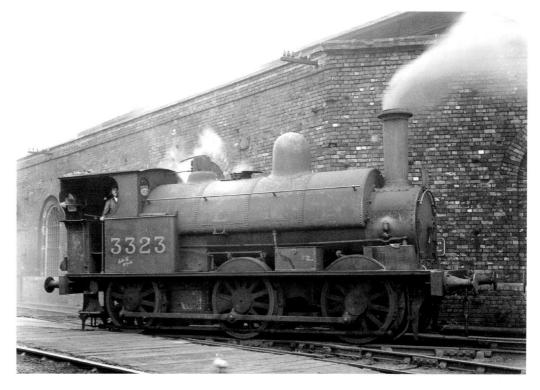

Sunday, 7 March 1948. Ex-LNWR Francis Webb-designed Class 5ft 6in (LMS Class 1P) 2-4-2 tank No. 6682 is seen here in Crewe Works yard after its withdrawal from service in December of the previous year. Constructed at Crewe during 1893, she would be 'motor fitted' in 1932.

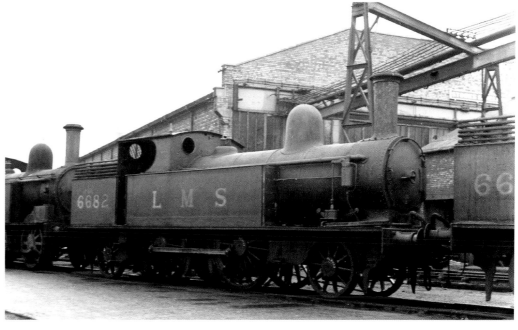

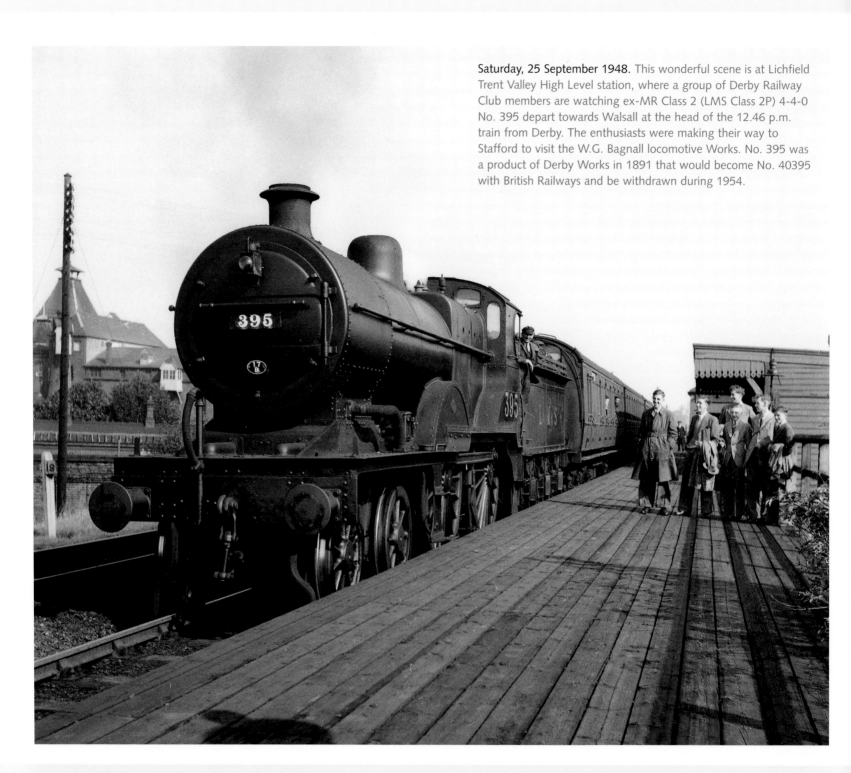

Saturday, 25 September 1948. This wonderful scene is at Lichfield Trent Valley High Level station, where a group of Derby Railway Club members are watching ex-MR Class 2 (LMS Class 2P) 4-4-0 No. 395 depart towards Walsall at the head of the 12.46 p.m. train from Derby. The enthusiasts were making their way to Stafford to visit the W.G. Bagnall locomotive Works. No. 395 was a product of Derby Works in 1891 that would become No. 40395 with British Railways and be withdrawn during 1954.

Sunday, 10 October 1948. Seen in Crewe Works yard after an overhaul, awaiting its new owner's identity but bearing its new number, is ex-LNWR 5ft 6in Class (LMS Class 1P) 2-4-2 tank No. 46666. Constructed at Crewe during 1893 and originally numbered 2146 by the LNWR, she would be withdrawn from service in 1954.

Sunday, 30 January 1949. A visit to Staveley shed on this day produced photographs of three 0-6-0 tanks. *Above:* Ex-LMS Class 3F No. 7627 was a 1928 product of William Beardmore & Co. in Glasgow that would be withdrawn during 1966, numbered 47627.

Samuel Johnson-designed ex-MR Class 1377 (LMS Class 1F) No. 1747 is seen in its original open-cab form. Initially constructed at Derby Works in 1884, she would be rebuilt during 1950 with a Belpaire firebox, but be withdrawn in 1954, numbered 41747.

Seen here in its rebuilt form with a Belpaire firebox, ex-MR Class 1337 (LMS Class 1F) No. 1753 was also a product of Derby Works during 1884. Rebuilt in 1926, she would be withdrawn in 1957, numbered 41753.

Sunday, 20 February 1949. Ex-LNWR Class 5ft 6in (LMS Class 1P) 2-4-2 tank No. 6635 is seen in Crewe Works yard. Constructed at Crewe during 1892, she would become number 46635 and be withdrawn in 1950.

Sunday, 20 February 1949. Also seen in Crewe Works yard is ex-LNWR 0-6-0 'Coal Engine' (LMS Class 2F) bearing her newly acquired number of 58343. Constructed at Crewe Works during 1881, she would give seventy-two years of service before being withdrawn in 1953.

Sunday, 20 February 1949. Bearing her new number but retaining her original owner's identity on the tender, ex-LMS Class 8F 2-8-0 No. 48347 is seen in the yard at Crewe Works. Constructed at Horwich Works during 1944, she would be withdrawn in 1967.

Saturday, 12 March 1949. Seen here in the ex-GWR Wolverhampton Works yard is ex-LNWR 0-6-0 'Coal Engine' (LMS Class 2F) No. 8108 minus its coupling rods. Constructed at Crewe Works in 1880, she would be sold to the Shropshire and Montgomeryshire Light Railway (S&MLR) during 1931, along with two of her classmates. Requisitioned by the War Department in 1941, she would be utilised until 1946, after which she was returned to British Railways stock in 1950 and subsequently scrapped. The S&MLR traced its ancestry back to 1866, being prematurely closed in 1880 with the line lying derelict afterwards. It was resurrected by Col. Stephens during 1911 and the passenger service lasted until 1933. The War Department took over the running of the line during 1941 to service the Central Armaments Depot at Nesscliffe, which was closed in 1960.

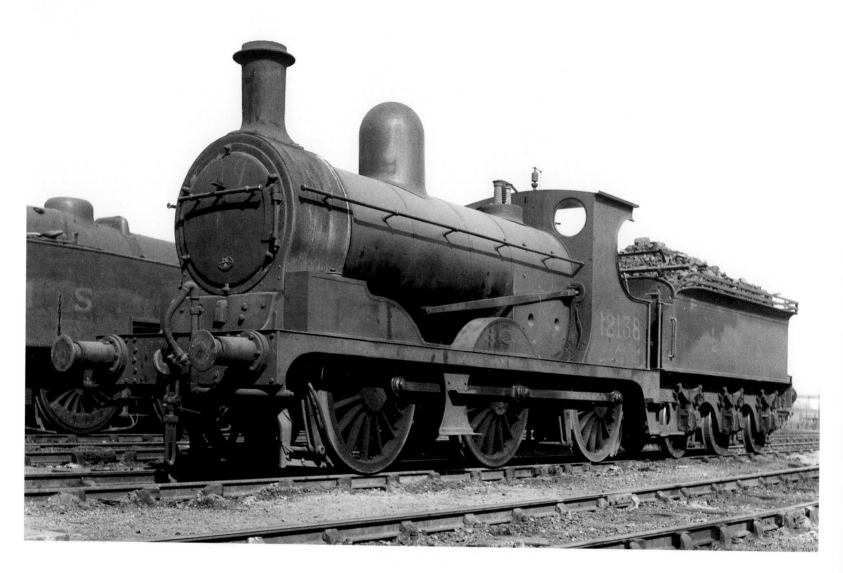

Sunday, 27 March 1949. Bearing the correct 26A shed code, ex-L&YR
Class 27 (LMS Class 3F) 0-6-0 No. 12138 is seen at Newton Heath shed.
Constructed at Horwich Works in 1891, she would become No. 52138 with
British Railways and be withdrawn during 1954.

Sunday, 27 March 1949. With her new owner's identity on the tender and bearing her new number, ex-L&YR Class 28 (LMS Class 3F) 0-6-0 No. 52582 is also seen at Newton Heath shed. A product of Horwich Works during 1893, she would be withdrawn in 1955.

Sunday, 27 March 1949. Sitting amongst a heap of scrap in Horwich Works yard, ex-L&YR Class 30 (LMS Class 6F) 0-8-0 No. 12834 is seen after being withdrawn from service earlier in March. She had been constructed at the same works during 1918.

Sunday, 27 March 1949. Two examples are seen here of John Aspinall's Class 5 (LMS Class 2P) 2-4-2 tanks for the L&YR. Constructed at Horwich Works during 1896, No. 10738 is in store at Newton Heath shed. She would be withdrawn from service later in 1949.

Sunday, 27 March 1949. Bearing its new British Railways number, but retaining its former owner's identity on the tank side, is No. 50642. Seen at Bolton shed, she was an earlier example of the class, entering service from Horwich Works in 1890. She was fitted for 'Push & Pull' working during 1934 and would be withdrawn from service in 1951.

Sunday, 27 March 1949. Here are two examples of John Aspinall's rebuilding of the earlier Class 528 William Barton Wright 0-6-0 tender locomotives into Class 23 0-6-0 saddle tanks for the L&YR. Withdrawn earlier in March, No. 11325 is seen in Horwich Works yard awaiting scrapping. Constructed by Sharp Stewart & Co. in 1877, she would be rebuilt as a saddle tank during 1892.

Sunday, 27 March 1949. With its new number, 51436 is seen here at Newton Heath shed. Constructed by Beyer Peacock & Co. during 1881, she would be rebuilt in 1895 and withdrawn during 1955 after seventy-four years of service.

Sunday, 27 March 1949. The William Stanier design of Class 3P 2-6-2 tanks introduced during 1935 were a taper-boiler version, with improved cab details of the earlier Henry Fowler parallel-boiler locomotives. Seen here at Horwich Works are classmates. *Top:* No. 179 was a product of Derby Works during 1938 that would be withdrawn in 1962. *Bottom:* No. 196 was a Crewe Works version also from 1938 that would also be withdrawn during 1962.

Sunday, 27 March 1949. Seen in Newton Heath shed yard, bearing a 24E Blackpool shed code, is ex-LMS Class 4P 'Compound' 4-4-0 No. 41195, looking splendid in its lined-out British Railways livery. An example of the final members of the class to be delivered from the Vulcan Foundry during 1927, she would be allocated to Blackpool for much of her working life and withdrawn after only thirty years of service in 1957.

Sunday, 27 March 1949. Waiting to be scrapped at Horwich Works are two of the ex-L&YR George Hughes-designed examples of his Class 8 four-cylinder 4-6-0s. *Top:* No. 10412 was constructed at Horwich during 1908 and would be rebuilt with Walschaerts valve gear and new cylinders in 1921, before being withdrawn in the month prior to this photograph. *Bottom:* No. 10432 was from a later batch that entered service from Horwich Works in 1922. She was withdrawn from service during the month of this photograph.

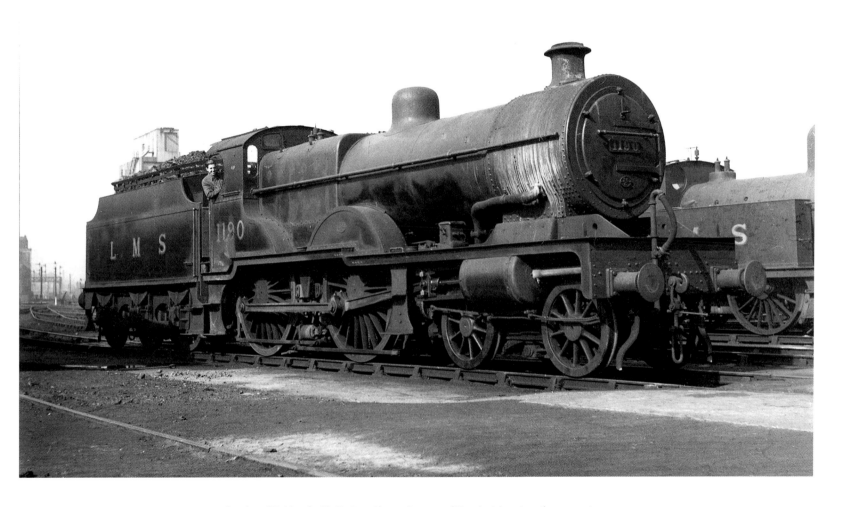

Sunday, 27 March 1949. In rather grimy condition but bearing the correct shed code, 26C ex-LMS Class 4P 4-4-0 'Compound' No. 1190 is seen at its home shed, Bolton. Constructed by the Vulcan Foundry in 1927, she would be numbered 41190 and withdrawn during 1958.

Sunday, 27 March 1949. At Newton Heath shed ex-LMS Class 5 2-6-0 No. 2867 is still wearing its previous owner's identity. Seen here bearing a 24F Fleetwood shed code, she was constructed at Crewe Works during 1930 and would be withdrawn in 1964.

Sunday, 27 March 1949. Standing in the yard at Horwich Works is Class 4MT 2-6-0 No. 43030, which has just completed construction and is waiting to enter service. Designed by Henry George Ivatt for the LMS and introduced during 1947, construction would continue until 1952 when 162 examples had been completed. The first fifty examples were fitted with double chimneys, which were later removed and replaced by single chimneys. No. 43030 would be withdrawn during 1966.

Sunday, 27 March 1949. Standing in Horwich Works yard and having just acquired her new British Railways number, ex-LMS Class 4F 0-6-0 No. 44594 was a product of Derby Works in 1939 that would be withdrawn from service during 1962. Note the particularly high-sided tender fitted to this locomotive.

Sunday, 27 March 1949. In Horwich Works yard is the impressive bulk of Henry Fowler-designed Class 7F 0-8-0 No. 9517, which had just been withdrawn from service during the same month. Constructed at Crewe Works during 1929, she gave only twenty years of service.

Saturday, 16 April 1949. Working hard with an 'up' mineral train and
seen passing Spondon station is ex-LMS Class 4F 0-6-0 No. 44556.
Bearing a 20H Lancaster shed code, she is a long way from home territory.
Constructed at Crewe Works during 1928, she would be withdrawn in 1963.

Tuesday, 3 May 1949. At the head of a short goods train ex-MR Class 4 (LMS Class 4F) 0-6-0 No. 43838 waits for the right of way at Chaddesdon sidings, Derby. A product of Derby Works in 1917, she would be withdrawn from service during 1956.

Tuesday, 3 May 1949. Also seen at Chaddesdon sidings is an example of the Matthew Kirtley-designed, double-framed Class 1 (LMS Class 2F) 0-6-0 for the Midland Railway. No. 58110, formerly No. 2630 with the LMS, had been constructed by Dübs & Co. in 1870 and she would survive eighty-one years to be withdrawn from service during 1951.

Sunday, 19 June 1949. Nearing completion of an overhaul, ex-LMS Class 5XP 'Patriot' 4-6-0 No. 45542 is seen in the yard at Crewe Works. Constructed at Crewe during 1934, she is bearing a 12B Carlisle Upperby shed code. An example of the unnamed members of the class, she would remain in this un-rebuilt, parallel-boiler form until she was withdrawn in 1962.

Sunday, 19 June 1949. Bearing a 1B Camden shed code and also nearing completion of an overhaul at Crewe Works is ex-LMS Class 6P 'Royal Scot' 4-6-0 No. 46151 *The Royal Horse Guardsman*. Constructed at Derby Works in 1930, she would be rebuilt with a taper boiler in 1953 and withdrawn during 1962.

Saturday, 2 July 1949. Seen at Birmingham New Street station on what the headlamp signifies is a local working heading in the 'down' direction, is ex-LMS Class 4P 'Compound' 4-4-0 No. 41116 piloting classmate No. 41111. The former was a product of Horwich Works during 1925 that would be withdrawn in 1957, and the latter was from Derby Works in 1925 and would be withdrawn from service during 1958.

Friday, 8 July 1949. Parked in the Derby Works yard is ex-S&DJR Class 7F 2-8-0 No. 53801. Constructed at the same works during 1914 for the Somerset and Dorset Joint Railway, she was numbered 81 by them before becoming No. 13801 with the LMS. She would be withdrawn during 1961, having given forty-seven years of service.

Sunday, 28 August 1949. At Trafford Park shed, ex-MR Class 2 (LMS Class 2P) 4-4-0 No. 40396 is simmering at the entrance. Constructed at Derby Works in 1891, she would be rebuilt with a Belpaire firebox during 1923 before being withdrawn in 1961.

Sunday, 28 August 1949. Ex-LMS Class 7F 0-8-0 No. 49532 is seen at Agecroft shed. Constructed at Crewe Works during 1929, she would give only twenty-seven years of service before being withdrawn in 1956.

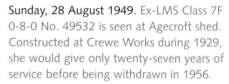

Saturday, 10 September 1949. Derby station pilot, ex-MR Class 2228 (LMS Class 1P) 0-4-4 tank No. 58077 has acquired its new number but is still bearing its previous owner's identity. Designed by Samuel Johnson and introduced in 1875, construction of the class continued until 1900, when 205 examples had entered service. No. 58077 was a product of Dübs & Co. in 1895 that would be rebuilt with a Belpaire firebox during 1927 and fitted for 'Push & Pull' operation at a later date. She would be withdrawn from service in 1955.

Friday, 16 September 1949. Sitting in the ex-GWR yard at Abergavenny station is ex-LNWR Class 0-6-2 'Coal Tank' No. 7752. Classified 2F by the LMS, all 300 examples of the class were constructed at Crewe Works with No. 7752 entering service during 1886. She was fitted with the LMS vacuum control system for 'Push & Pull' working in 1927 and would be withdrawn from service during 1952.

Friday, 23 September 1949. Seen here at Derby, preparing to shunt some wagons, is ex-L&YR Class 21 (LMS Class 0F) 0-4-0 saddle tank No. 51235. Constructed at Horwich Works in 1906 and numbered 840 by that railway, she would become No. 11235 with the LMS and be withdrawn after fifty-two years of service in 1958.

Sunday, 5 February 1950. Bearing a 7B Bangor shed code, ex-LMS Class 2P 2-6-2 tank No. 41200 is seen in Crewe Works yard. Designed by Henry George Ivatt for the LMS prior to the nationalisation of the railways, only ten examples of the class entered service during LMS days. The class consisted of 130 examples, the final ten coming from Derby Works and the remainder all coming out of Crewe Works. No. 41200 entered service from Crewe Works late in 1946 and would be withdrawn during 1965.

Sunday, 5 February 1950. Scheduled to be withdrawn during this month, ex-LNWR Class G1 (LMS Class 6F) 0-8-0 No. 49059 awaits its fate in Crewe Works yard. Originally constructed at the same works in 1898 to a design by Francis Webb as a Class A, three-cylinder compound, she would be rebuilt as a two-cylinder Class D locomotive during 1908 and finally to a Class G1 in 1928.

Sunday, 5 February 1950. The Charles Bowen-Cooke-designed Class G1 (LMS Class 7F) 0-8-0 No. 49279, which was constructed at the same works during 1918, is also awaiting scrapping.

Sunday, 5 February 1950. In Crewe Works yard a veteran locomotive can be seen: ex-LNWR 0-6-0 'Coal Engine' (LMS Class 2F) No. 58328 was another product of Crewe Works, entering service during 1875. Numbered 203 by the LNWR, she would become 8115 with the LMS and give seventy-eight years' service before withdrawal in 1953.

Sunday, 5 February 1950. Destined to be the last of its class, Francis Webb's 0-4-2 saddle tank No. 47862 was an example of the twenty locomotives constructed at Crewe Works between 1896 and 1901 for the LNWR as dock or yard shunters. Utilising a Bissel trailing truck, which enabled them to negotiate sharp curves, they were classified 1F by the LMS with No. 47862 spending its last days as one of the Crewe Works shunters. She would be withdrawn from service during 1956.

Sunday, 5 February 1950. At Crewe North shed, ex-MR Class 2 (LMS Class 2P) 4-4-0 No. 527 is still sporting its previous owner's identity. Constructed at Derby Works during 1898, she would be rebuilt with a Belpaire firebox as seen here in 1913 and withdrawn from service during 1956, numbered 40527.

Sunday, 5 February 1950. Also seen at Crewe North shed is ex-LMS Class 5 4-6-0 No. 44807, one of the few members of the class constructed at Derby Works. Entering service during 1944, she would be withdrawn in 1968. Introduced during 1934, the 'Black Staniers', or 'Black 5s' as they became known, extended to 842 examples with the last entering service in 1951.

Sunday, 5 February 1950. In Crewe Works yard, Class 2F 2-6-0 No. 46433
is bearing a 1A Willesden shed code. Constructed at Crewe in 1948,
she would spend her last days allocated to Carnforth shed before being
withdrawn during 1967.

Sunday, 5 February 1950. The 4-6-0s of Class '19in Goods' for the LNWR were designed by George Whale and introduced during 1906, with construction continuing until 1909 when 170 examples had entered service. Seen here is No. 8824, the last of all the ex-LNWR 4-6-0s to be withdrawn. Constructed at the same works during 1908, this photograph shows her in Crewe Works yard during the month of her withdrawal.

Sunday, 5 February 1950. William Stanier's first class of 'Pacific' design for the LMS was his 'Princess Royal' Class 7P, of which a total of twelve were constructed at Crewe Works during 1933 and 1935. Seen here in Crewe Works yard, No. 46200 *The Princess Royal* was the first example to enter service in July 1933 and gave less than thirty years of service before withdrawal in 1962.

Sunday, 5 February 1950. In Crewe Works yard. This rear-view angle gives some impression of the size of the coal bunker on the William Stanier designed ex-LMS Class 4P 2-6-4 tanks. With a load of 3.5 tons and a 2,000-gallon water capacity, these were ideal locomotives for suburban passenger traffic and were utilised throughout the old LMS territory. No. M2641 was constructed at Derby Works during 1938 and would be withdrawn in 1962.

Sunday, 5 February 1950. Also in Crewe Works yard is Ex-LMS Class 6P 'Royal Scot' 4-6-0 No. 46165 *The Ranger 12th London Regt*, constructed at Derby Works during 1930. She would be rebuilt with a taper boiler in 1952. Destined to be something of a wanderer, she was allocated to many sheds throughout her working life, from Upperby in Carlisle to Holyhead and Crewe North. She would be withdrawn in 1964.

Sunday, 5 February 1950. Again in Crewe Works yard, ex-LMS Class 5XP 'Patriot' 4-6-0 No. 45506 *The Royal Pioneer Corps* was a product of Crewe Works in 1932 that would acquire its name in 1948. She would give thirty years of service before being withdrawn in the un-rebuilt form seen here during 1962.

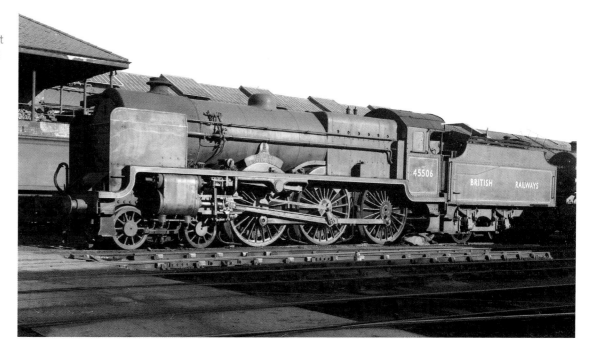

Friday, 17 February 1950. Ex-MR Class 2 (LMS Class 2P) 4-4-0 No. 40513 is seen here laying a dense smokescreen as she departs from Derby Midland at the head of the 3.38 p.m. local working to Birmingham New Street.

Constructed by Sharp Stewart & Co. during 1899, she would be rebuilt with a Belpaire firebox in 1912 and withdrawn during 1959, having given sixty years of service.

Thursday, 16 March 1950. In sparkling condition, ex-MR Class 2 (LMS Class 2P) 4-4-0 No. 40383 is passing Derby South Junction with the Engineers' Inspection Saloon. Constructed at Derby Works in 1888, she would be rebuilt with a Belpaire Firebox during 1909 and withdrawn after sixty-four years of service in 1952.

Thursday, 16 March 1950. At Chaddesden sidings in Derby, ex-LMS Class 3F 0-6-0 tank No. 47436 is in very clean condition and probably ex-works. Constructed by Hunslet & Co. in Leeds during 1927, she would be withdrawn in 1960.

Monday, 3 April 1950. This wonderful scene sees ex-MR Class 2 (LMS Class 2P) 4-4-0 No. 40370 about to pass the Way & Works Sidings signal box in Derby at the head of the 11.15 a.m. from Derby Midland to Nottingham Midland local working. Constructed at Derby Works during 1886 and rebuilt with a Belpaire firebox in 1922, she would be withdrawn during 1951 after sixty-five years' service.

Friday, 28 April 1950. Seen in Derby Works yard is the first member of the ex-LMS Class 4P three-cylinder 2-6-4 tanks designed by William Stanier and constructed at the same works during 1934. These were specifically designed for use on the London, Tilbury and Southend passenger traffic. No. 42500 survived withdrawal in 1962 to become part of the National Collection.

Wednesday, 10 May 1950. In grimy looking condition but making a spirited departure from Derby Midland station, ex-LMS Class 5 2-6-0 No. 42822 is at the head of the 1.50 p.m. parcels working to Bristol. Constructed at Horwich Works in 1929, she was fitted with Lentz rotary cam poppet valve gear during 1932 and finally equipped with Reidinger rotary cam valve gear in 1952. She would be withdrawn ten years later during 1962.

Wednesday, 10 May 1950. Seen departing Derby Midland at the head of the 3.38 p.m. working to Birmingham New Street local is Class 2F 2-6-0 No. 46454. A product of Crewe Works during 1950, she would give only sixteen years of service before being withdrawn in 1966.

Wednesday, 10 May 1950. Still bearing her former owner's identity and number is ex-MR Class 2 (LMS Class 2P) 4-4-0 No. 426. Constructed at Derby Works in 1896, where she can be seen here, she would be fitted with a Belpaire firebox during 1916 and withdrawn in 1957.

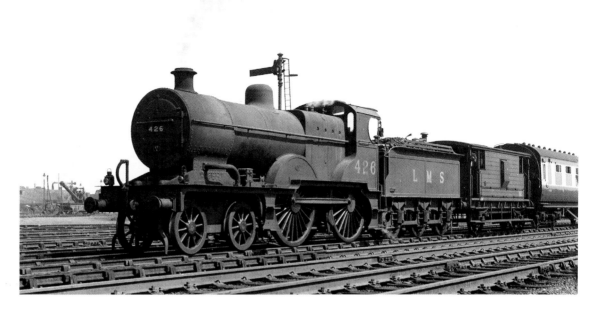

Thursday, 11 May 1950. Seen here at Derby Works, looking very smart in her new livery, is ex-LMS Class 3P 2-6-2 tank No. 40006. Designed by Henry Fowler and introduced during 1930, a total of seventy examples were constructed, all at Derby Works, with the last example entering service during 1932. Several members of the class were equipped with condensing gear to enable them to work the London suburban traffic, while others were fitted for 'Push and Pull' operation. No. 40006 was from a 1930-constructed batch that would be withdrawn in 1962.

Saturday, 20 May 1950. This fine photograph shows ex-MR Class 2 (LMS Class 2P) 4-4-0 No. 40472 departing from Derby Midland station at the head of the 5.12 p.m. local working to Sheffield. Allocated to 18C Hasland shed in Chesterfield, she was a product of Derby Works in 1895 that would be fitted with a Belpaire firebox during 1923 and withdrawn from service in 1955.

Monday, 22 May 1950. Standing over an inspection pit in Derby shed yard is ex-LMS Class 2P 4-4-0 No. 40633 with an interesting history. She was constructed at Derby Works during 1928 and became No. 44 with the Somerset and Dorset Joint Railway. Transferred to LMS stock in 1930, she became No. 633 with them, fitted with the Dabeg feed water heater in 1933 before being withdrawn during 1959. The Dabeg Automatic Locomotive Feed Pump was a patented apparatus to improve the efficiency of locomotives so fitted. By preheating the water feed with the use of exhaust steam and using a mechanically fitted pump system, it was claimed that more water evaporation was gained when compared to conventional water-injection methods, and therefore water and coal consumption were reduced.

Thursday, 29 June 1950. With the striking tower of St Andrew's Church in the background, ex-MR Class 2 (LMS Class 2P) 4-4-0 No. 40535 is seen passing the Way & Works Sidings signal box at the head of the 11.15 a.m. Derby Midland to Nottingham Midland working. Bearing a 16A Nottingham shed code, she had been constructed at Derby Works in 1899 and fitted with a Belpaire firebox during 1913. She would be withdrawn in 1955.

Thursday, 29 June 1950. The locomotive seen here in Derby Works yard has just been withdrawn after giving seventy-four years of service. Ex-MR Class 1252 (LMS Class 1P) 0-4-4 tank No. 58036 had been constructed by Neilson & Co. during 1876 and originally numbered 1281 by the MR, later becoming No. 1255.

Thursday, 29 June 1950. Another resident of Derby shed, 17A, was Class 4 2-6-0 No. M3010. She was constructed at Horwich Works during 1948 and incorporated a double chimney, which would later be replaced by a single version. She would be withdrawn in 1967.

Friday, 30 June 1950. Seen departing from Derby Midland station at the head of the 5.15 p.m. working to Worcester Shrub Hill station is ex-MR Class 4 (LMS Class 4P) 'Compound' 4-4-0 No. 41012. Constructed at Derby Works during 1905, she would be rebuilt with a superheating boiler in 1921 and withdrawn in 1951.

Friday, 30 June 1950. Arriving at Derby Midland station at the head of the 11.40 a.m. working from Bournemouth is ex-LMS Class 4P 'Compound' 4-4-0 No. 41047. Constructed at Derby Works in 1924, she would be withdrawn thirty years later during 1954.

Saturday, 8 July 1950. At Birmingham New Street station, ex-LNWR 0-6-2 'Coal Tank' (LMS Class 2F) No. 58928 is working as the station pilot. Constructed at Crewe Works during 1888, she was originally numbered 1068 by the LNWR and later became 7803 with the LMS. She would give sixty-three years' service before being withdrawn in 1951.

Thursday, 13 July 1950. Constructed by Kitson & Co. of Leeds during 1932, the first five examples of a Class 0F 0-4-0 saddle-tank locomotive for the LMS were originally numbered 1540 to 1544, later becoming 7000 to 7004. Seen here at Derby in ex-works condition is No. 47000. She spent much of her working life on the Cromford and High Peak line before being withdrawn in 1966.

Saturday, 15 July 1950. Ex-LMS Class 3P 2-6-2 tank No. 40118 is taking water at Chinley whilst working the 6.05 p.m. Manchester Central to Buxton local. An example of the William Stanier version of the earlier Henry Fowler design, she was constructed at Derby Works during 1935 and would be withdrawn in 1961.

Friday, 28 July 1950. Seen near Pear Tree and Normanton station at the head of the 5.15 p.m. Derby Midland to Worcester Shrub Hill working is ex-MR Class 4 (LMS Class 4P) 4-4-0 'Compound' No. 41012. This appeared to be a regular turn for this locomotive.

Tuesday, 29 August 1950. Ex-MR Class 4 (LMS Class 4F) 0-6-0 No. M4010, seen here minus its tender, is in Derby Works yard prior to an overhaul. She had been constructed at the same works during 1921 and was allocated to Staveley (Barrow Hill) at the time.

Friday, 22 September 1950. The same locomotive is looking splendid after its overhaul and repainting with her new number, 44010, and a 21A Saltley shed code. She would be withdrawn from service during 1963.

Wednesday, 4 October 1950. Looking in splendid external condition at Derby shed, ex-MR Class 2228 (LMS Class 1P) 0-4-4 tank No. 58090 is still bearing her previous owner's identity. Constructed by Dübs & Co. during 1900 and fitted with a Belpaire firebox in 1927, she would be withdrawn in 1953.

Thursday, 15 March 1951. Constructed at Derby Works during 1896, ex-MR Class 2 (LMS Class 3F) 0-6-0 No. 43194 originally worked for the S&DJR as their No. 62 until 1930, when she was absorbed into LMS stock. Seen here at Derby Works, bearing an appropriate 71H Templecombe shed code and equipped with a single-line tablet exchange mechanism, she would be withdrawn during 1960.

Saturday, 21 April 1951. A comparison here of two examples of the Henry Fowler-designed ex-LMS Class 4F 0-6-0s, both seen near Duffield north of Derby. Seen at the head of an 'up' goods, No. 44092 was a product of Kerr Stuart & Co. during 1925 that would be withdrawn in 1964.

Saturday, 21 April 1951. Also seen at the head of an 'up' goods is No. 44587, which was constructed at Derby Works in 1939 and withdrawn during 1965. Note for comparison the tenders on these locomotives: No. 44587 is paired with a high-sided version.

Tuesday, 24 April 1951. Ex-L&YR Class 21 (LMS Class 0F) 0-4-0 saddle tank
No. 51217, in rather grimy condition, is seen in Derby shed yard. Constructed
at Horwich Works during 1895 and originally numbered 1290 by the L&YR,
she would become No. 11217 with the LMS and be withdrawn in 1961.

Tuesday, 24 April 1951. The ex-LNWR super-power seen here in Derby shed yard is Class G2A (LMS Class 7F) 0-8-0 No. 48944. Originally designed by Francis Webb as a class of four-cylinder locomotives, the Class G2As were introduced during 1901, with No. 48944 coming out of Crewe Works in 1903 numbered 1308 by that company. She would be rebuilt as a class G1 two-cylinder locomotive in 1924 and would undergo further rebuilding during 1940 as a class G2A utilising a higher boiler pressure. She would be withdrawn in 1957.

Thursday, 10 May 1951. More ex-LNWR superpower seen here, this time at Preston shed. Class G2A (LMS Class 7F) 0-8-0 No. 49104 was originally constructed as a class G two-cylinder locomotive at Crewe Works in 1910 and rebuilt as a class G1 during 1924 before being finally rebuilt as a class G2A incorporating a higher boiler pressure in 1940. She would be withdrawn from service in 1962.

Thursday, 31 May 1951. Entering service barely two months prior to this photograph, Class 5 4-6-0 No. 44686 is seen here making a fine spectacle whilst departing from Derby Midland station with the 11.15 a.m. working to Crewe. Constructed at Horwich Works and entering service in April 1951, she was one of only two examples of the class fitted with a revised form of Caprotti valve gear utilising outside drive, roller bearings on all axles and a double chimney. She would be withdrawn during 1965.

Thursday, 31 May 1951. Another of William Stanier's successful designs and in beautifully clean condition, ex-LMS Class 5XP 'Jubilee' 4-6-0 No. 45626 *Seychelles* is departing from Derby Midland station at the head of the 12.45 p.m. Newcastle to Bristol working. Constructed at Crewe Works during 1934, she would be withdrawn in 1965.

Tuesday, 5 June 1951. In ex-works condition ex-LMS Class 4F 0-6-0 No. 44055, bearing a 22B Gloucester, Barnwood shed code, is seen passing Derby South Junction at the head of a train of empty mineral wagons. Constructed at Derby Works during 1925, she would be withdrawn from service in 1963.

Wednesday, 6 June 1951. Ex-MR Class 2 (LMS Class 3F) 0-6-0 No. 43709 is seen at Chaddesden Sidings, Derby. Constructed by Sharp Stewart & Co. in 1901, she would give sixty-one years' service and be withdrawn during 1962.

Friday, 6 July 1951. Ex-S&DJR Class 7F 2-8-0 No. 53800 is in ex-works condition at Derby. Constructed at Derby during 1914, she appears to have spent the bulk of her working life allocated to Bath Green Park shed, 22C.

Numbered 80 by the S&DJR, she would become No. 9670 and later No. 13800 with the LMS and be withdrawn during 1959.

Monday, 16 July 1951. On the former Cromford and High Peak Railway, Ex-LNWR Class 2228 (LMS Class 1P) 2-4-0 tank No. 58092 is seen at Black Rocks with a train of Sheep Pasture to Middleton empty mineral wagons. Designed by Francis Webb, this class of passenger tanks were all constructed at Crewe Works between 1876 and 1880. No. 58092 entered service during 1877 and would give seventy-five years of service before being withdrawn in 1952.

Monday, 16 July 1951. In Cromford station yard ex-NLR Class 75 (LMS Class 2F) 0-6-0 tank No. 58862 is shunting wagons. Constructed at the North London Railway Bow Works during 1879, a total of thirty examples of this dock-shunting tank to a design by J.C. Park entered service. No. 58862 would give seventy-seven years' service and be withdrawn in 1956.

Monday, 16 July 1951. *Seen at the head of a long train of empty mineral wagons near High Peak station is ex-LMS Class 8F 2-8-0 No. 48552. A product of Darlington Works during 1945, she would be withdrawn in 1967, having given only twenty-two years' service.*

Monday, 16 July 1951. Ex-LMS 'Garratt' 2-6-0+0-6-2 Beyer Garratt No. 47989 is working a 'down' empty mineral wagon train at Matlock station. Weighing in at a massive 152.5 tons and constructed by Beyer Peacock & Co. in Manchester during 1930, a total of thirty-three examples entered service, three during 1927 and the remainder in 1930. No. 47989 would give only twenty-five years' service before being withdrawn in 1955.

Tuesday, 17 July 1951. Seen passing Lichfield station on the 'down' fast road at the head of the 11.50 a.m. London Euston to Workington 'Lakes Express' is ex-LMS Class 6P 4-6-0 'Rebuilt Royal Scot' No. 46141 *The North Staffordshire Regiment.* Constructed by the NBL in Glasgow with a parallel boiler during 1927, she would be rebuilt with a taper boiler in 1950, as seen here, and withdrawn in 1964.

Tuesday, 17 July 1951. The crew member of ex-MR Class 1134A (LMS Class 0F) 0-4-0 Saddle tank No. 41523 is looking relaxed during a break in shunting duties in Burton-on-Trent yard. Designed by Samuel Johnson and constructed at Derby Works during 1903, No. 41523 would be withdrawn in 1955.

Thursday, 19 July 1951. More than three years after nationalisation, ex-MR Class 2 (LMS Class 2P) 4-4-0 No. 404 is still wearing its former owner's identity. Bearing a 17A Derby shed code, she had been constructed by Sharp Stewart & Co. in 1892 and rebuilt with a Belpaire firebox during 1918. She would be withdrawn after sixty-five years of service in 1957.

Thursday, 19 July 1951. Seen arriving at Chaddesden sidings in Derby at the head of a goods train is ex-LMS Class 5XP 4-6-0 'Jubilee' No. 45611 *Hong Kong*. Bearing a 16A Nottingham shed code, she was constructed at Crewe Works during 1934 and would be withdrawn thirty years later in 1964.

Opposite: Friday, 20 July 1951. It is approaching 4.00 p.m. and ex-LMS Class 4P 'Compound' 4-4-0 No. 40927 is making a spectacular exit from Derby Midland station whilst piloting ex-LMS Class 5 4-6-0 No. 44855 at the head of the northbound 'The Devonian'. The 'Compound' was a product of the Vulcan Foundry in 1927 that would be withdrawn during 1957 while the 'Black 5' entered service from Crewe Works in 1944 and would be withdrawn during 1967. In the background is Engine Sidings No. 1 signal box.

Right: Saturday, 21 July 1951. Seen near Spondon is ex-LMS Class 5 2-6-0 No. 42768 at the head of a special working from Scarborough to Sawley Junction. Constructed at Crewe Works during 1927, she would be withdrawn in 1963.

Saturday, 21 July 1951. Bearing a 27B Aintree shed code, ex-LMS Class 5 2-6-0 No. 42728 is at the head of a Blackpool to Sawley Junction working, seen here departing from Derby Midland station. A product of Horwich Works that entered service during 1927, she would be withdrawn in 1963.

Opposite: Saturday, 28 July 1951. Standing under the sweeping curve of the roof at Birmingham New Street station is ex-LMS Class 5 2-6-0 No. 42818. Constructed at Horwich Works during 1929 and numbered 13118 by the LMS, she would later become No. 2828 with them. During 1931 she was fitted with the Lentz rotary cam poppet valve gear as seen here. In 1953 this would be replaced by Reidinger rotary cam valve gear. Seen here bearing a 21A Saltley shed code, she would be withdrawn in 1962.

Above: Friday, 28 December 1951. Class 5 4-6-0 No. 44744, wearing a 22A Bristol, Barrow Road shed code, is seen in Derby shed yard. She was constructed during 1948 at Crewe Works as part of a batch of twenty examples of this class which were fitted with Caprotti valve gear in an attempt to reduce maintenance costs. She would be withdrawn after only fifteen years' service in 1963.

Tuesday, 18 March 1952. Being utilised as a Derby Works shunter, ex-L&YR Class 21 (LMS Class 0F) 0-4-0 saddle tank No. 51235 is looking very smart wearing her new British Railways livery.

Tuesday, 25 March 1952. Seen here at Derby Works is No. 58100, specifically designed and constructed by the Midland Railway to work as a banking locomotive on the Lickey incline. Initially numbered 2290, she entered service in 1919 and would become No. 22290 with the LMS and then 58100 with British Railways. Constructed at Derby Works, the four-cylinder 0-10-0 continued working until 1956 when she was replaced by other locomotives.

Sunday, 20 April 1952. Bearing a 17B Burton shed code, Class 2MT 2-6-0 No. 46494 is standing in Derby shed yard. An example of the Darlington Works' constructed locomotives of this class, she entered service during 1951 and would be withdrawn only eleven years later in 1962.

Friday, 9 May 1952. With Derby Station North signal box in the background, ex-MR Class 2 (LMS Class 2P) 4-4-0 No. 40337, bearing an 18C Hasland shed code, is seen departing with the 5.12 p.m. local to Sheffield Midland. Constructed at Derby Works in 1882, she would be rebuilt on several occasions, at one point receiving a Belpaire firebox. Giving seventy-six years of service, she would be withdrawn in 1958. Note the first vehicle behind the locomotive, which is marked 'Kitchen Car'.

Monday, 12 May 1952. Near Spondon Junction, east of Derby, ex-MR Class 2 (LMS Class 2P) 4-4-0 No. 40556 is seen working the 5.15 p.m. Nottingham to Derby local. Constructed by Neilson Reid & Co. during 1901, she would be rebuilt with a Belpaire firebox in 1913 and withdrawn during 1956.

Friday, 16 May 1952. Un-rebuilt ex-LMS Class 5XP 'Patriot' 4-6-0 No. 45509 *The Derbyshire Yeomanry* is seen departing from Derby Midland station at the head of the 12.45 p.m. Newcastle to Bristol express. Constructed at Crewe Works during 1932, she would acquire her name in 1951 and be withdrawn in 1961.

Friday, 4 July 1952. More than four years after nationalisation, ex-MR Class 2 (LMS Class 2F) 0-6-0 No. 2998 still retains its LMS identity. Constructed by Neilson & Co. during 1876, she is seen at Derby Works awaiting an overhaul. She would emerge later carrying the number 58171 and would be withdrawn after eighty-three years' service in 1959.

Saturday, 26 July 1952. At the head of the ten-coach 11.40 a.m. Yarmouth Beach to Derby Midland working, ex-LMS Class 4F 0-6-0 No. 44401 is seen near Spondon Junction. Constructed by the NBL in Glasgow during 1927, she would be withdrawn in 1965.

Tuesday, 5 August 1952. At Chesterfield, Midland shed is seen ex-S&DJR Class 3F 0-6-0 No. 43211 bearing an 18C Hasland shed code. Constructed by Derby Works during 1896, she was numbered 66 by the S&DJR, before becoming 3211 with the LMS. She would be withdrawn from service during 1961.

Thursday, 14 August 1952. Seen here passing Spondon Junction is ex-MR Class 2 (LMS Class 3F) 0-6-0 No. 43273 at the head of the 5.37 p.m. Spondon to Darley Dale local consisting of eleven coaches. Constructed by Neilson & Co. during 1891, she would give sixty-four years of service and be withdrawn in 1955.

Saturday, 10 October 1952. Ex-MR Class 2228 (LMS Class 1P) 0-4-4 tank No. 58076 is waiting to depart from Great Malvern station with the 10.23 a.m. working to Ashchurch. Constructed by Dübs & Co. in 1895, she would be rebuilt with a Belpaire firebox in 1928 and withdrawn during 1953.

Wednesday, 22 April 1953. Standing in Derby Works yard is ex-MR Class 2 (LMS Class 2F) 0-6-0 No. 58308. Constructed by Neilson Reid & Co. during 1901, she would be withdrawn from service in 1959.

Wednesday, 22 April 1953. Seen at Derby shed still bearing her former owner's identity but wearing her new British Railways number, ex-LMS Class 4F 0-6-0 No. 44142 was a product of Crewe Works in 1925 that would be withdrawn during 1959. Note that the tender has been constructed with covers to enable operation with snow-clearing duties.

Wednesday, 22 April 1953. Sitting in Derby Works yard minus her coupling rods is ex-NLR Class 75 (LMS Class 2F) 0-6-0 tank No. 58858. Constructed at the Bow Works of the North London Railway during 1888, she would be withdrawn from service in the month following this photograph.

Wednesday, 22 April 1953. Arriving at Derby Midland station with the 3.05 p.m. working from Lincoln is ex-LMS Class 4P 'Compound' 4-4-0 No. 41060. Looking in good external condition, she had been constructed at Derby Works during 1924 and would be withdrawn in 1958.

Saturday, 25 April 1953. Standing in Duffield station at the head of the Stephenson Locomotive Society/Manchester Locomotive Society High Peak Rail Tour is ex-MR Class 2228 (LMS Class 1P) 0-4-4 tank No. 58077. Constructed by Dübs & Co. during 1895, she would be withdrawn in 1955.

Saturday, 25 April 1953. At Middleton Top, ex-NLR Class 75 (LMS Class 2F) 0-6-0 tank No. 58860 is waiting to depart with the High Peak Rail Tour. Constructed at the Bow Works of the NLR during 1892, she would be withdrawn from service in 1957.

Monday, 29 June 1953. In Derby shed yard ex-LNWR Class G2A (LMS Class 7F) 0-8-0 No. 49153 is going through a disposal procedure, having had the smoke box cleared of char and the fire cleaned.

Constructed at Crewe Works during 1902 as a Class B four-cylinder locomotive, she would be rebuilt on several occasions, finally becoming a Class G2A locomotive in 1940, and be withdrawn from service during 1959.

Monday, 14 June 1954. Standing in Derby Works yard, after an overhaul that saw her original larger boiler replaced by a smaller version, is ex-LMS Class 7F 2-8-0 No. 53807. Constructed by Robert Stephenson & Co. during 1925 and numbered 87 by the S&DJR, she would become No. 9677 with the LMS, and later No. 13807. She would be withdrawn from service in 1964.

Saturday, 19 June 1954. Seen here at speed passing Spondon station, Class 5 4-6-0 No. 44661 at the head of the 12.05 p.m. Derby Midland to London St Pancras express. Constructed at Crewe Works in 1949, she would give only eighteen years of service and be withdrawn during 1967. The 'Black 5s' were constructed over a period of seventeen years from 1934 until 1951 by Crewe, Derby and Horwich Works, with two outside contractors, Armstrong Whitworth Ltd and the Vulcan Foundry Ltd, also supplying many examples.

Monday, 30 August 1954. Sitting in the yard at Aintree shed, Liverpool is
ex-L&YR Class 23 (LMS Class 2F) 0-6-0 saddle tank No. 51460. Originally
constructed by the Vulcan Foundry during 1881 as a William Barton Wright
Class 528 0-6-0 tender locomotive, she was rebuilt in 1896 as a saddle tank.
She would be withdrawn from service during the month of this photograph.

Monday, 30 August 1954. Also in the yard at Aintree is ex-L&YR
Class 27 (LMS Class 3F) 0-6-0 No. 52311 bearing a 27B Aintree
shed code. Constructed at Horwich Works during 1895, she would be
withdrawn in 1962, having given sixty-seven years of service.

Monday, 30 August 1954. The Henry Fowler-designed Class 7F 0-8-0s for the LMS introduced during 1929 were intended to be an updated version of the former LNWR 0-8-0s. Incorporating a Belpaire firebox with an increased boiler pressure of 200psi, their drawback was the use of Midland Railway-designed axle boxes, which were prone to failure. Seen here at Aintree shed on this day were:

Left: No. 49659, which was constructed at Crewe Works in 1932 and would be withdrawn during 1957. *Below left:* No. 49566 was another product of Crewe Works from an earlier 1929 batch, she would also be withdrawn in 1957. *Opposite:* No. 49586 also entered service from Crewe Works during 1929 but would be withdrawn thirty years later in 1959.

Monday, 30 August 1954. Seen in Aintree shed yard, ex-L&YR Class 23 (LMS Class 2F) 0-6-0 saddle tank No. 51413 had originally been constructed by Beyer Peacock & Co. in 1881 as a Class 528 0-6-0 tender locomotive. Rebuilt as a saddle tank during 1895, she would give a total of eighty years' service before being withdrawn in 1961.

Monday, 30 August 1954. Ex-L&YR Class 24 (LMS Class 2F) 0-6-0 tank No. 51535 was an example of John Aspinall's design of shunting locomotive that was introduced in 1897 with a total of twenty class members, all from Horwich Works. She was numbered 1355 by the L&YR, later becoming No. 11535 with the LMS, and was withdrawn during 1956.

Monday, 30 August 1954. Destined to be withdrawn from service during the month after this photograph, at Aintree shed is ex-L&YR Class 27 (LMS Class 3F) 0-6-0 No. 52299. She had been constructed at Horwich Works during 1895.

Monday, 30 August 1954. Ex-LMS Class 7F 0-8-0 No. 49511 has just been coaled up at Aintree shed. A product of Crewe Works in 1929, she would be withdrawn during 1959.

Monday, 30 August 1954. Seen at Aintree on hump-shunting duties, ex-LMS Class 7F 0-8-0 No. 49505 was constructed at Crewe Works during 1929 and would be withdrawn in 1960.

Friday, 17 September 1954. In Derby Works yard, ex-MR Class 4
(LMS Class 4F) 0-6-0 No. 44557 is bearing a 71H Templecombe shed code.
Originally constructed by Armstrong Whitworth & Co. during 1922 and
numbered 57 by the S&DJR, she would be absorbed into LMS stock in
1930 and withdrawn during 1962.

Friday, 17 September 1954. Seen arriving at Derby Midland station with the 11.40 a.m. express from Bournemouth is ex-LMS Class 4P 'Compound' 4-4-0 No. 40934, bearing a 22B Gloucester, Barnwood shed code. Constructed by the Vulcan Foundry in 1927, she would be withdrawn thirty years later during 1957.

Wednesday, 29 September 1954. Approaching Derby South Junction is ex-MR Class 2 (LMS Cass 2P) 4-4-0 No. 40418, which is hauling an Officers Saloon Special. Constructed by Sharp Stewart & Co. during 1892, she would be rebuilt with a Belpaire firebox in 1914 and be withdrawn after sixty-five years of service during 1957.

Sunday, 28 November 1954. Bearing a 5A Crewe North shed code, ex-LMS Class 5XP 'Patriot' 4-6-0 No. 45510 is seen in Crewe Works yard. Constructed at the same works in 1932, she would be something of a wanderer, being allocated to sheds as far apart as Willesden and Carlisle Upperby throughout her working life of thirty years, before being withdrawn during 1962.

Sunday, 28 November 1954. Standing in Crewe Works yard in very clean condition is ex-LMS Class 4P 2-6-4 tank No. 42493. Constructed at Derby Works during 1937, she would be withdrawn in 1964.

Sunday, 28 November 1954. An example of possibly the finest of William Stanier's locomotive designs is seen here in Crewe North shed yard. Ex-LMS Class 8P 'Princess Coronation' 4-6-2 No. 46242 *City of Glasgow* was a product of Crewe Works during 1940, entering service as a streamlined version of the class, although the streamlining would be removed in 1947. She would at various times be allocated to Camden, Polmadie and Crewe North sheds and would be withdrawn in 1963.

Sunday, 28 November 1954. Also seen in Crewe North shed yard is ex-LMS Class 5XP 'Rebuilt Patriot' 4-6-0 No. 45512 *Bunsen*. Originally constructed at Crewe Works during 1932 with a parallel boiler, she would be rebuilt with a taper boiler as seen here in 1948 and withdrawn from service during 1965.

Sunday, 28 November 1954. Ex-LMS Class 4P 'Compound' 4-4-0 No. 41157 is seen here under the coaling stage at Crewe North shed. Constructed by the NBL in Glasgow during 1925, she would be withdrawn in 1960.

Opposite: Saturday, 30 April 1955. Seen after arrival at Manchester Exchange station, the 9.10 a.m. service from Liverpool Lime Street has a beautifully clean ex-LMS Class 5XP 'Rebuilt Patriot' 4-6-0 at its head. No. 45531 *Sir Frederick Harrison* was originally constructed at Crewe Works during 1933 and then rebuilt with a taper boiler, as seen here in 1947. She would be withdrawn during 1965 with the station not lasting much longer: it was closed in 1969 and later demolished.

Top right: Wednesday, 1 June 1955. On 23 May 1955 the ASLEF union called a strike which lasted until 14 June of the same year. Many services were curtailed and goods traffic mounted up in yards and docks. The NUR footplate crews continued working, and seen here arriving at Derby with 'strike special' No. 22, the 2.00 p.m. Leeds to London St Pancras, is ex-LMS Class 5 No. 45274. Constructed by Armstrong Whitworth Ltd during 1936, she would be withdrawn in 1967.

Bottom right: Friday, 8 July 1955. Witnessed by a lone enthusiast at the platform end, another arrival at Derby Midland station is seen here with ex-LMS Class 5XP 'Jubilee' 4-6-0 No. 45572 *Eire* at the head of the 2.15 p.m. Bristol to York express. Constructed by the NBL in Glasgow during 1934, she was initially named *Irish Free State,* but this was changed to *Eire* in 1938. She would give thirty years of service before being withdrawn in 1964.

Saturday, 30 July 1955. The Windermere portion of the 'Lakes Express' is preparing to depart from that station behind ex-LMS Class 6P 'Rebuilt Royal Scot' 4-6-0 No. 46100 *Royal Scot*. Bearing a 1B Camden shed code, she had originally been constructed as LMS No. 6152 *Kings Dragoon Guardsman* at Derby Works during 1930, and swapped identities with the original LMS No. 6100 in 1933 prior to her departure to America to visit the Chicago Century of Progress Exhibition. Returning to the United Kingdom later that year, she would be rebuilt in 1950 with a taper boiler and withdrawn in 1962. She would avoid the scrapheap, being purchased and kept at the Bressingham Steam Museum for many years before being overhauled and returned to main-line steam work during 2016.

Wednesday, 10 August 1955. In ex-works condition after an overhaul and seen in Derby Works yard is ex-LMS Class 7F 2-8-0 No. 53806. Constructed by Robert Stephenson & Co. during 1925 as part of a second batch of this class for the S&DJR, she was originally fitted with a large-diameter boiler, which would be exchanged for a smaller-diameter example during this overhaul. Originally numbered 86 by the S&DJR, she would become No. 9676 and later No. 13806 with the LMS. She would be withdrawn during 1964.

Friday, 23 December 1955. Seen departing from Derby Midland station at the head of the 8.05 a.m. Newcastle to Cardiff relief working is ex-LMS Class 5 4-6-0 No. 44814, bearing a 21A Saltley shed code. Constructed during 1944 at Derby Works, she would be withdrawn in 1967.

Friday, 23 December 1955. Bearing a 15A Wellingborough shed code, and seen at the head of the 1.08 p.m. Derby Midland to Lincoln St Marks working, is Class 4 2-6-0 No. 43040. Originally constructed with a double chimney at Horwich Works in 1949, this would be replaced with a single version during 1953 and she would be withdrawn from service in 1966.

Friday, 23 December 1955. Passing the former 'ticket platform' just outside Derby Midland station, ex-MR Class 2 (LMS Class 2P) 4-4-0 No. 40416 is working the 12.45 p.m. local to Birmingham via Lichfield, bearing a 17A Derby shed code. She had been constructed by Sharp Stewart & Co. in 1892 and would be rebuilt with a Belpaire firebox during 1944. Having given sixty-seven years of service, she would be withdrawn in 1959.

Saturday, 31 December 1955. Carrying 'The Devonian' headboard, ex-LMS Class 5 4-6-0 No. 44826 is showing signs of leaking glands as she departs from Derby Midland station with this westbound train. Constructed at Crewe Works during 1944, she would be withdrawn in 1967. This titled train had been introduced during 1927 as a joint working between the LMS and the GWR connecting Bradford and Leeds with Paignton in Devon via Sheffield, Derby, Birmingham, Cheltenham and Bristol. The service would be withdrawn in 1975.

Saturday, 7 April 1956. In Derby Works yard is ex-LMS Class 3F 0-6-0 tank No. 47316, formerly numbered 25 by the S&DJR. Constructed by W.G. Bagnall & Co. of Stafford during 1929, she is still bearing a 71G Bath shed code. She would be withdrawn in 1962.

Friday, 20 April 1956. Making a spirited start from Derby Midland station whilst piloting 'Black 5' No. 44856 at the head of the 12.40 p.m. Newcastle to Bristol express is ex-LMS Class 2P 4-4-0 No. 40682. Constructed at Derby Works during 1932, she would be withdrawn only twenty-nine years later in 1961. The 'Black 5' was a product of Crewe Works during 1944 that would be withdrawn in 1967.

Friday, 20 April 1956. Seen in Derby Works yard is ex-LMS Class 2P 4-4-0 No. 40633. Constructed at Derbyduring 1928 and entering service as No. 44 with the S&DJR, she was absorbed into LMS stock in 1930 and fitted with the Dabeg feed waterheater system during 1933. She would be withdrawn in 1959 after only thirty-one years' service.

Friday, 4 May 1956. Designed by Henry Fowler specifically for shunting in docks and yards with tight curves, the Class 2F 0-6-0 tanks consisted of only ten examples, all constructed at Derby Works during 1928 and 1929. No. 47166 is seen here at her birthplace wearing a 6F Bidston, Birkenhead shed code. An example of the 1928 batch, she would variously be allocated to Bidston, Edge Hill and Birkenhead sheds before being withdrawn in 1963.

Friday, 11 May 1956. Seen here in Derby Works yard is ex-LNWR Class G2A (LMS Class 7F) 0-8-0 No. 49268 looking splendid after an overhaul. She had been constructed at Crewe Works during 1917 as a class G1 locomotive that was later converted to a class G2A. Bearing a 10A Springs Branch, Wigan shed code, she would be withdrawn in 1959.

Friday, 11 May 1956. Also in Derby Works yard is ex-LMS Class 4F 0-6-0 No. 44282 bearing a 20G Hellifield shed code. Her tender is equipped with a weather shield to protect the crew in adverse conditions. Constructed at Derby in 1926, she would be withdrawn during 1963.

Tuesday, 11 September 1956. Pausing to take water at Sheep Pasture on the former Cromford and High Peak Railway, ex-LMS Class 0F 0-4-0 saddle tank No. 47000 has three water carriers in tow. Constructed by Kitson & Co. in 1932, she would spend the bulk of her working life on the C&HPR, allocated to 17D Rowsley shed. She would be withdrawn from service during 1966.

Above: **Thursday, 25 October 1956.** With large lumps of coal piled high into her cab-side bunker, ex-L&YR Class 21 (LMS Class 0F) 0-4-0 saddle tank No. 51217 is seen at Derby Works. Constructed at Horwich Works during 1895, she would be withdrawn in 1961.

Opposite: **Thursday, 28 March 1957.** In beautifully clean condition and bearing the correct 'four lamp' head code, ex-LMS Class 5 4-6-0 No. 45447 is seen at Derby shed piloting classmate No. 45444 prior to working the Royal Train on that day. Both locomotives were constructed by Armstrong Whitworth Ltd during 1937, and both would be withdrawn at the cessation of main-line steam operations in the United Kingdom in August 1968.

Saturday, 13 April 1957. A visit to the ex-GCR lines west of Chester brought the opportunity to view passenger workings on the Wrexham to Seacombe route that wandered through the Wirral.

Opposite: The 11.27 a.m. working from Seacombe to Wrexham Central is seen passing the signal box at Hawarden Bridge Junction as it approaches Shotton High Level station. In charge is ex-LMS Class 3P 2-6-2 tank No. 40202, which had been constructed at Crewe Works during 1938 and would be withdrawn in 1962. *Right:* Class 2P 2-6-2 tank No. 41244 is seen arriving at Buckley Junction station with the 11.25 a.m. Wrexham Central to Seacombe train. Constructed at Crewe Works during 1949, she would be withdrawn in 1966.
Below right: Waiting to depart from Seacombe station at the head of the 2.29 p.m. working to Wrexham Central is ex-LMS Class 3P 2-6-2 tank No. 40083. Constructed at Derby Works in 1935, she would be withdrawn during 1962.

Tuesday, 30 April 1957. Seen passing the Way & Works Sidings signal box at the head of the 12.12 p.m. Derby Midland to Nottingham Midland local working is ex-MR Class 2 (LMS Class 2P) 4-4-0 No. 40411. Constructed by Sharp Stewart & Co. during 1892, she would be withdrawn after sixty-nine years of service in 1961. Note the vehicle directly behind the locomotive tender: it is a twelve-wheeled 'sleeping car'. The distinctive church spire in the background is St Andrews, known locally as the railway church.

Tuesday, 4 June 1957. Slowing for the stop at Warrington, ex-LMS Class 8P 'Princess Coronation' 4-6-2 No. 46230 *Duchess of Buccleuch* is at the head of the 10.05 a.m. Glasgow Central to Birmingham New Street express. Constructed during 1938 as a non-streamlined version of the class and allocated to Polmadie shed in Glasgow for the bulk of her working life, she would be withdrawn in 1963.

Friday, 26 July 1957. Seen arriving at Derby Midland station at the head of the 11.27 a.m. express form Bournemouth West is ex-LMS Class 4P 'Compound' 4-4-0 No. 41095. Constructed at Derby Works in 1925, she would be withdrawn during 1958.

Saturday, 28 September 1957. This day saw the running of the Talyllyn Railway Preservation Society 'Talyllyn AGM Special' from London Paddington to Towyn via Shrewsbury, with the section from Shrewsbury to Towyn being handled by ex-L&YR Class 5 (LMS Class 2P) 2-4-2 tank No. 50781 piloting ex-GWR Class 3200 'Earl' 4-4-0 No. 9021. Seen here departing from Moat Lane Junction on the outward leg, No. 50781 later failed and another locomotive substituted for the return working. No. 50781 was constructed at Horwich Works during 1897 and would be withdrawn in 1960.

Saturday, 28 September 1957. At Moat Lane Junction, Class 2MT 2-6-0
No. 46522 is preparing to depart with the 2.45 p.m. stopper to Brecon via
Builth Wells. Constructed as late as 1953 at Swindon Works, she would serve
for just fourteen years before being withdrawn during 1967.

Saturday, 28 September 1957. At Moat Lane Junction on the same day was classmate No. 46519, also constructed at Swindon Works in 1953, seen here bearing an 89B Croes Newydd shed code. She would be withdrawn during 1966 after only thirteen years' service.

Wednesday, 6 November 1957. At St Helens shed ex-L&YR Class 27 (LMS Class 3F) 0-6-0 No. 52125 is sitting in the yard. Constructed at Horwich Works during 1891, she would be withdrawn during the month of this photograph.

Saturday, 11 January 1958. Bearing a 71G Bath shed code, ex-S&DJR
Class 7F 2-8-0 No. 53801 is seen in Derby Works yard. Constructed at the
same works during 1914 and numbered 81 by the S&DJR, she would be
withdrawn in 1961.

Opposite: Saturday, 31 May 1958. Emerging from the lingering smoky gloom whilst approaching Stockport station, ex-LMS Class 5 4-6-0 No. 45434 is at the head of the 12.10 p.m. Birmingham New Street to Manchester London Road express. Constructed by Armstrong Whitworth Ltd during 1937, she would be withdrawn in 1966.

Tuesday, 3 June 1958. Ex-S&DJR Class 7F 2-8-0 No. 53800 is seen in Derby shed yard, bearing an 82F Bath, Green Park shed code. Constructed at the same works in 1914, she was numbered 80 by the S&DJR and would be withdrawn during 1959.

Wednesday, 4 June 1958. Seen approaching Stockport station at the head of the 2.00 p.m. Manchester London Road to London Euston express is ex-LMS Class 6P 'Rebuilt Royal Scot' 4-6-0 No. 46122 *Royal Ulster Rifleman*. Originally constructed by the NBL in Glasgow during 1927, she would be rebuilt with a taper boiler in 1945 and withdrawn during 1964.

Wednesday, 4 June 1958. This busy scene at the north end of Crewe station sees ex-LMS Class 8P 'Princess Coronation' 4-6-2 No. 46225 *Duchess of Gloucester* bearing 'The Mid-Day Scot' headboard waiting to take over that train from classmate No. 46231 *Duchess of Atholl*. Both locomotives had been constructed at Crewe Works during 1938, No. 46225 as a streamlined example of the class and No. 46231 without streamlining. The former would be withdrawn in 1964, with the latter going two years earlier in 1962.

Monday, 7 July 1958. Ex-LMS Class 5XP 'Jubilee' 4-6-0 No. 45616 *Malta GC* is seen departing from Derby Midland station at the head of 'The Palatine'. Constructed at Crewe Works during 1934 and originally named *Malta* – the name changed after the island was awarded the George Cross – she would be withdrawn during 1961. 'The Palatine' London St Pancras to Manchester service had been inaugurated by the LMS in 1938, but was suspended during the war years and only reinstated in 1957.

Thursday, 10 July 1958. With a full head of steam but some leaking glands, ex-LMS Class 6P 'Rebuilt Royal Scot' 4-6-0 No. 46152 *The Kings Dragoon Guardsman* is seen about to pass the Way & Works Sidings signal box with the 10.25 a.m. Manchester Central to London St Pancras working. Originally constructed by the NBL in Glasgow during 1927 and entering service as No. 6100 *Royal Scot,* she would switch identities with No. 6152 in 1933, that locomotive going to the Chicago Century of Progress Exhibition bearing the *Royal Scot* name and number. The locomotive seen here was rebuilt with a taper boiler in 1945 and withdrawn from service during 1965.

Saturday, 20 September 1958. This day saw the running of the Stephenson Locomotive Society (Midland Area) Tanat Valley and Llanfyllin Rail Tour, commencing at Gobowen. The tour visited the remaining part of the Tanat Valley line as far as Llanrhaiadr Mochnant and the old Cambrian Railway line to Llanfyllin. Class 2MT 2-6-0 No. 46509, bearing an 89A Shrewsbury shed code, was the motive power. No. 46509 was constructed at Swindon Works during 1952 and would be withdrawn in 1966.

Saturday, 27 September 1958. Seen at Brecon station is Class 2MT 2-6-0 No. 46520, which had been constructed at Swindon Works in 1953 and would be withdrawn during 1967, after only fourteen years' service.

Thursday, 13 November 1958. Standing in Derby Works yard minus its coupling rods is ex-L&YR Class 21 (LMS Class 0F) 0-4-0 saddle tank No. 51232. Constructed at Horwich Works during 1906, she would be withdrawn in 1963.

Wednesday, 18 February 1959. Waiting to depart from Chester General station at the head of an express is ex-LMS Class 5XP 'Jubilee' 4-6-0 No. 45674 *Duncan*. Constructed at Crewe Works during 1935, she would be withdrawn in 1964.

Thursday, 13 August 1959. Seen in Derby Works yard after completion of an overhaul is ex-LMS Class 8F 2-8-0 No. 48199, bearing a 41E Staveley, Barrow Hill shed code. Constructed by the NBL in Glasgow during 1942, she would be withdrawn in 1967.

Friday, 12 May 1961. Seen departing from Birmingham New Street station, at the head of the 1.40 p.m. working to Liverpool Lime Street, is ex-LMS Class 5 4-6-0 No. 45146. Constructed by Armstrong Whitworth Ltd in 1935, she would be withdrawn thirty years later during 1965.

Monday, 7 August 1961. With holidaymakers crowding the platform at
Rugeley Town station, the 8.32 a.m. Dudley Port to Llandudno special
working has ex-LMS Class 5XP 'Rebuilt Patriot' 4-6-0 No. 45529 *Stephenson*
at its head. Originally constructed at Crewe Works during 1933, she would be
rebuilt with a taper boiler during 1947 and then withdrawn in 1964.

Sunday, 25 March 1962. Seen at Kingmoor shed in Carlisle is ex-LMS Class 8P 'Princess Coronation' 4-6-2 No. 46223 *Princess Alice*. Constructed at Crewe Works during 1937 as a streamlined example of the class, the streamlining would be removed in 1946. She would spend the bulk of her working life allocated to Polmadie shed in Glasgow and would be withdrawn during 1963.

Tuesday, 17 July 1962. Ex-LMS Class 6P 'Rebuilt Royal Scot' 4-6-0 No. 46161 *King's Own* is seen arriving at Birmingham New Street station at the head of the 8.45 a.m. London Euston to Wolverhampton High Level working. Bearing a 5A Crewe North shed code, she had been constructed during 1930 at Derby Works, rebuilt with a taper boiler in 1946 and would be withdrawn during 1962.

Above: Saturday, 8 September 1962. This day saw the Stephenson Locomotive Society/Manchester Locomotive Society Leicestershire Rail Tour departing form Manchester Piccadilly, visiting Burton-on-Trent and a large number of destinations in Leicestershire behind a variety of locomotives. Seen arriving at Burton-on-Trent is ex-LMS Class 4P 2-6-4 tank No. 42343: constructed at Derby Works in 1929, she would be withdrawn during 1965.

Opposite: Saturday, 1 June 1963. Working hard with a long train of mineral wagons, ex-LMS Class 8F 2-8-0 No. 48519 is seen moving in the 'up' direction at Danesmoor, near Clay Cross. Constructed as part of a Railway Executive order at Doncaster Works during 1944, she would continue in service until the end of main-line steam in the United Kingdom before being withdrawn in August 1968. The impressive church tower seen on the hill in the background belongs to St Lawrence in North Wingfield.

Friday, 20 March 1964. Approaching Castle Bromwich station at the head of an 'up' goods is Class 4 2-6-0 No. 43122. Constructed at Horwich Works during 1951, she would be withdrawn from service in 1967.

Friday, 5 June 1964. At Bordesley Station on the ex-GWR line near Birmingham New Street, ex-LMS Class 8F 2-8-0 No. 48436 is at the head of a long goods train, which contains a large number of cattle wagons. Constructed as part of a Railway Executive order at Swindon Works in 1944, she would be withdrawn during 1966.

Above: Wednesday, 14 October 1964. Drifting through Nuneaton, Trent Valley station with a long goods train is ex-LMS Class 5 4-6-0 No. 45398. Constructed during 1937 by Armstrong Whitworth Ltd, she would be withdrawn in 1965.

Overleaf: Friday, 2 April 1965. Seen at Water Orton, near Birmingham, at the head of a 'down' goods is ex-LMS Class 8F 2-8-0 No. 48176, which had been constructed by the NBL in Glasgow during 1942 and would be withdrawn in 1967.